# .EV Battery Supply Chain: Sustainable EV Batteries and Ethical Sourcing for Batteries

All rights reserved. No part of this publication may be reproduced, distributed, or transmitted in any form or by any means, including photocopying, recording, or other electronic or mechanical methods, without the prior written permission of the publisher, except in the case of brief quotations embodied in critical reviews and certain other noncommercial uses permitted by copyright law.

Copyright ©Michael Kante, 2022.

TABLE OF CONTENT

CHAPTER 1            The Importance of Raw Materials in EV Batteries

CHAPTER 2            Critical Raw Materials for EV Batteries

CHAPTER 3            Sustainability and Ethical Sourcing

CHAPTER 4            Supply Chain Dynamics and Challenges

CHAPTER 5            Charging Infrastructure and Grid Integration

CHAPTER 6            Solid-State Batteries

CHAPTER 7            Environmental Impacts and Sustainability of EV Batteries

CHAPTER 8            Sand Performance of EV Batteries

CHAPTER 9            Economic and Market Dynamics of EV Batteries

# CHAPTER 1

## The Importance of Raw Materials in EV Batteries

The anticipated growth in battery manufacturing might be hampered by geopolitical unrest as well as the fragile and unpredictable character of the crucial raw-material supply chain. This would hinder the adoption of electric vehicles (EVs) by the general public and the transition to an electrified future.

Suppliers and original equipment manufacturers are facing a threat to their profit margins as a result of the skyrocketing costs of crucial battery metals, as seen in the figure below from S&P Global Commodity Insights. According to recent research conducted by S&P Global Mobility Auto Supply Chain & Technology Group, this scenario has swiftly resulted in a rise in the pricing of both components and vehicles.

The growth of the raw materials supply chain across markets is also being impacted by trade friction and environmental, social, and governance issues. As a whole, these changes contribute to the difficulties that are associated with the shift to electric vehicles.

It will be necessary for a rapidly expanding industry to have a steep growth curve in order to achieve its volume targets. In order for original equipment manufacturers (OEMs) to achieve their goals of selling BEVs and hybrids, S&P Global Mobility anticipates that the industry will need around 3.4 terawatt hours (TWh) of lithium-ion batteries on an annual basis by the year 2030. The medium- and heavy-duty, as well as the micro-mobility, as well as consumer electronics, and the growing need for stationary energy storage, are not included in this number. The 2021 production for the car industry: 0.29 TWh.

There is no such thing as elements like lithium, nickel, and cobalt just materializing out of thin air and transforming into electric vehicle batteries and other components. The development chain is extensive and intricate, from their difficulty to extract to their complicated refining. The intermediary phases between excavation and final assembly represent a specific choke point in terms of skill and market presence. Currently, China is the obvious leader in materials refinement, as well as the packaging and manufacturing of battery cells. At issue is which other governments will step forward to aid this sector transition.

In terms of getting battery raw materials, the issue comes down to Who needs what, where will it come from, who will provide it, and who is best poised to gain from this growing dependence on a handful of crucial elements?

The current S&P Global Mobility analysis assesses the battery raw material supply chain from extraction to vehicle, identifying:

A number of unknown firms that will play a crucial part in the processing and development of battery-electric vehicle (BEV) technologies that will underlie the light passenger cars of the future decade and beyond;

Potential trade friction might imply difficulty for major car firms in extricating themselves from an established, agile, and cost-effective supply of processed materials originating from or via mainland China;

Some OEMs are seeking the value and confidence of "locked in" supply chain connections from mine to vehicle, minimizing the dependence on unpredictable spot markets and/or a requirement to collaborate with less established industry partners.

Overview of EV Battery Components

Cathodes:

The positive electrode in a battery where the reduction reaction happens when discharging.

Common materials include Lithium Cobalt Oxide (LiCoO2), Lithium Nickel Manganese Cobalt Oxide (NMC), Lithium Iron Phosphate (LFP), and Lithium Nickel Cobalt Aluminum Oxide (NCA).

These materials determine the battery's capacity, energy density, and thermal stability.

Anodes:

The negative electrode where oxidation occurs after discharging.

Most often produced from graphite because to its stability and effectiveness in intercalating lithium ions.

Research is continuing into silicon anodes, which offer larger capacity but encounter issues with expansion while charging.

Electrolytes:

The medium that facilitates ionic conductivity between the cathode and anode.

Typically a lithium salt (such as LiPF6) is dissolved in a combination of organic solvents.

Solid-state electrolytes are a developing technology that might increase safety and energy density by removing flammable liquid components.

Separators:

Thin membranes are used between the cathode and anode to avoid physical contact while permitting ion passage.

Made from polymers like polypropylene or polyethylene.

Essential for safety, since they prevent short circuits and thermal runaway.

Role of Raw Materials in Battery Performance and Longevity

Lithium:

Key to attaining high energy density, which translates to extended driving range for EVs.

Essential for sustaining cycle life, since it supports numerous charge-discharge cycles.

High lithium concentration may prevent battery deterioration and boost lifetime.

Cobalt:

Provides thermal stability, which is critical for safety, particularly under heavy loads and quick charging.

Enhances energy density but is pricey and related with supply chain and ethical problems.

Reduction of cobalt content is a target area to increase sustainability and minimize costs.

Nickel:

Contributes greatly to energy density, allowing for bigger-capacity batteries.

Less costly than cobalt, although the processing and extraction are ecologically problematic.

Higher nickel content batteries (like NMC 811) are becoming increasingly common, albeit they need more rigorous thermal control.

Graphite:

Provides a stable framework for lithium-ion intercalation.

Contributes to the efficiency and rate capabilities of the battery, altering charge times and performance.

Research investigating alternatives, such as silicon or graphene, attempts to enhance capacity and minimize size.

Importance of Sustainable Sourcing

Environmental Impact:

Lithium extraction may lead to water depletion and pollution, particularly in dry places like the Lithium Triangle in South America.

Cobalt mining in the DRC typically leads in deforestation, habitat damage, and pollution.

Nickel mining and processing are energy-intensive and may contribute to large carbon emissions and trash creation.

Ethical Concerns:

Cobalt mining in the DRC is related to grave human rights violations, including child labor and dangerous working conditions.

Ensuring ethically obtained materials needs openness and conformity to international labor regulations.

Companies are being forced to check their supply chains and guarantee ethical sourcing procedures.

Economic Viability:

Sustainable sourcing helps assure long-term supply by limiting resource depletion and decreasing geopolitical concerns.

Investment in recycling and alternative materials may ameliorate raw material shortages and lessen dependence on volatile areas.

Establishing secure and ethical supply chains promotes brand reputation and compliance with regulatory norms.

Overview of the EV Battery Supply Chain

Stages of the Supply Chain

Mining:

The extraction of raw materials from the soil is the initial stage in the supply chain.

Major sources:

Lithium: Extracted from brine pools (Argentina, Chile) and hard rock mining (Australia).

Cobalt: Predominantly mined in the Democratic Republic of Congo (DRC).

Nickel: Extracted in nations including Indonesia, the Philippines, and Russia.

Graphite: Sourced from both natural sources (China, Brazil) and synthetic manufacturing.

Refining:

Raw materials are treated to remove and purify the useful constituents.

This level is energy-intensive and needs extensive infrastructure:

Lithium: Processed to generate lithium carbonate or lithium hydroxide.

Cobalt: Refined to generate cobalt sulfate for battery production.

Nickel: Processed into high-purity nickel sulfate for use in batteries.

Graphite: Purified and shaped for use in anodes.

Manufacturing:

Involves the fabrication of battery cells and packs.

Key processes:

Cathode and Anode Production: Involves mixing, coating, and drying of materials.

Cell Assembly: Stacking or winding of electrodes, adding electrolytes, and closing cells.

Battery Pack Assembly: Integrating cells into modules and packs, including battery management systems (BMS).

Recycling:

The end-of-life stage is when materials are retrieved from old batteries.

Methods include

Mechanical Recycling: Crushing and separating battery components.

Hydrometallurgical Processes: Using aqueous solutions to extract metals.

Pyrometallurgical Processes: High-temperature smelting to recover metals.

Recycling is vital for sustainability, lowering the demand for new raw materials and limiting environmental impact.

Key Players in the Supply Chain

Mining Companies:

Examples: Albemarle (lithium), Glencore (cobalt), Vale (nickel).

These corporations spend extensively in extraction technology and sustainability initiatives.

Battery Manufacturers:

Examples: Panasonic, LG Chem, CATL, Tesla.

Responsible for creating and manufacturing cells, typically cooperating with automakers.

Automakers:

Examples: Tesla, General Motors, Nissan, Volkswagen.

They integrate batteries into EVs, focusing on vehicle performance, range, and safety.

Recycling Firms:

Examples: Redwood Materials, Umicore.

These firms specialize in extracting valuable materials from used batteries to reintegrate them into the supply chain.

Global Distribution of Raw Materials and Their Impact on the Supply Chain

Lithium:

Major deposits are in the Lithium Triangle (Chile, Argentina, Bolivia), Australia, and China.

Distribution affects global trade, with countries like China dominating lithium processing and production.

Cobalt:

Over 60% of the world's cobalt comes from the DRC.

This concentration poses risks due to political instability and ethical concerns in the region.

Nickel:

Major producers include Indonesia, the Philippines, Russia, and Canada.

Nickel supply is vulnerable to geopolitical tensions and environmental regulations.

Graphite:

Natural graphite production is dominated by China, with significant contributions from Brazil and India.

Synthetic graphite production involves complex processes often controlled by companies in the U.S., Japan, and Europe.

The global distribution of these materials means that any disruption, whether political, environmental, or economic, can have significant ripple effects throughout the supply chain. Understanding these dynamics is crucial for managing risk and ensuring a stable supply of critical materials for EV battery production.

This thorough introduction sets the foundation for the subsequent chapters by offering a full knowledge of the significance of raw materials, the complexity of the supply chain, and the difficulties and possibilities in sustainable sourcing. Each part gives a thorough dive into the essential issues that drive EV battery production and the larger industry. Let me know if you want to continue with the next chapter or any particular aspect you want to explore more!

# CHAPTER 2

## Critical Raw Materials for EV Batteries

Expected development of electric vehicle (EV) sales has led to worries about sourcing mineral inputs utilized in EV batteries. Various nations and corporations have expressed plans to expedite the adoption of EVs in the transportation industry. Such governmental and private obligations show that

EV sales might continue into the projected future, with some predictions predicting 200 million

total EVs sold by 2030. More than 16 million total EVs have been sold globally, with roughly

6.6 million EVs sold in 2021. The U.S. EV market is tiny when compared to those in China and

Europe: new U.S. EV registrations were somewhat less than 10% of new worldwide EV registrations in

2021, whereas registrations in China were 50% of the world total and European registrations were 35%.

As the bulk of EV manufacture and sales occur outside the United States, so does the majority of EV battery

production. While China accounts for over 70% of global EV battery production capacity, the United States has developed battery supply networks for part of its requirements. China's supremacy in EV battery manufacture is akin to its dominance in

mining and extraction of the minerals required in EV batteries. The potential for a quicker worldwide shift to EVs leads

some to doubt the domestic availability of the minerals and resources for the home fabrication of EV batteries.

Currently, lithium-ion batteries are the leading kind of rechargeable batteries used in EVs. The most widely used

varieties are lithium cobalt oxide (LCO), lithium manganese oxide (LMO), lithium iron phosphate (LFP), lithium nickel cobalt aluminum oxide (NCA), and lithium nickel manganese cobalt oxide (NMC). Graphite is currently widely used as the

anode in lithium-ion batteries. These EV battery chemistries depend on five critical minerals whose domestic supply is

potentially at risk for disruption: lithium, cobalt, manganese, nickel, and graphite. The U.S. Geological Survey designated

these and other minerals as "critical," according to the methodology codified in the Energy Act of 2020.

The United States is heavily dependent on imports for these minerals for use in EV batteries and other applications. The

United States currently mines some lithium, cobalt, and nickel, but it does not currently mine any manganese or graphite.

Various companies have indicated plans to expand the mineral production of these minerals. Recycling products containing

these minerals contribute to some domestic production, and it represents further potential contributions to domestic supply.

Additional research to increase EV battery efficiencies or into new battery chemistries can reduce the requirements of these

critical minerals for EV battery production.

The 117th Congress has considered and may choose to consider further, various options related to EV adoption and enhanced domestic production of minerals used in EV batteries. Of the options considered, some have been included in enacted

legislation. The Infrastructure Investment and Jobs Act (IIJA, P.L. 117-58) includes multiple sections related to EV adoption

and enhancing the domestic supply of the critical minerals used in EV batteries. Some examples include Section 11401, Grants

for Charging and Fueling Infrastructure; Section 40201, Earth Mapping Resources Initiative; Section 40207, Battery

Processing and Manufacturing; Section 40208, Electric Drive Vehicle Battery Recycling and Second-Life Applications

Program; Section 40210, Critical Minerals Mining and Recycling Research; Section 40401, Department of Energy Loan

Programs; Section 71101, Clean School Bus Program; Division J, and Title VIII, National Electric Vehicle Formula Program.

In addition to ongoing federal programs related to EV batteries and changes resulting from provisions in the IIJA, Congress

could consider further changes related to the domestic supply of critical minerals used in EV batteries. Some additional

related areas include mining on federal lands, taxes and tariffs, and EV battery chemistry research, among others.

Lithium

Sources:

Brine Pools:

Found in salt flats, particularly in the Lithium Triangle (Chile, Argentina, Bolivia).

Extraction involves pumping brine to the surface and allowing it to evaporate, leaving lithium-rich salts.

Advantages: Lower production cost and lower environmental impact compared to hard rock mining.

Disadvantages: Slow process, significant water use, and potential impact on local water tables.

Hard Rock Mining:

Extracted from spodumene ores, especially in Australia.

Involves standard mining procedures followed by crushing and chemical processing to extract lithium.

Advantages: Higher lithium concentration and quicker extraction procedure.

Disadvantages: Higher expense and larger environmental effect compared to brine extraction.

Clay:

Emerging source in locations like Nevada, USA.

Lithium is recovered from clay deposits using several chemical methods.

Advantages: Potentially enormous reserves, closer to main markets.

Disadvantages: Still in development, with greater extraction costs and technological obstacles.

Mining Processes:

Brine Evaporation:

Brine is poured into enormous evaporation ponds.

Sunlight evaporates the water over 12-18 months, leaving lithium-rich salts.

The salts are then treated to make lithium carbonate or lithium hydroxide.

Hard Rock Mining:

Ore is mined and crushed.

Crushed ore is subsequently treated using flotation to concentrate spodumene.

Concentrated spodumene is further treated with sulfuric acid to generate lithium sulfate, which is subsequently turned to lithium hydroxide or carbonate.

Direct Extraction:

Newer technology targeted at speeding up the extraction process.

Methods include ion-exchange, solvent extraction, and electrochemical procedures.

Advantages: Faster and perhaps more ecologically friendly, although yet under research and scale.

Market Dynamics:

Supply and Demand:

Lithium demand is driven by the fast expansion of the EV industry and energy storage technologies.

Supply is concentrated in a few nations, leading to significant geopolitical issues.

Pricing Trends:

Prices are variable, driven by market demand, production rates, and geopolitical concerns.

Investments in new mining projects and extraction technology attempt to stabilize and enhance supply.

Future Outlook:

Continuous advancement in extraction technology.

Increased recycling initiatives to extract lithium from discarded batteries.

Potential for additional sources (e.g., geothermal brines) to diversify supply.

Cobalt

Ethical Sourcing:

Challenges in the Democratic Republic of Congo (DRC):

Over 60% of world cobalt supply originates from the DRC.

Issues include child labor, dangerous working conditions, and political instability.

Artisanal mining (small-scale, typically informal mining) is ubiquitous, leading to substantial social and environmental consequences.

Efforts to Improve Conditions:

Initiatives like the Fair Cobalt Alliance and the Responsible Cobalt Initiative strive to improve mining operations and working conditions.

Companies are increasingly buying cobalt via approved supply chains to guarantee ethical operations.

Mining Conditions:

Artisanal vs. Industrial Mining:

Artisanal mining: Small-scale, labor-intensive, generally uncontrolled.

Industrial mining: Large-scale activities with more strict controls and monitoring.

Industrial miners tend to have superior safety and environmental procedures but face criticism for the displacement of local people.

Alternative Sources and Reduction Strategies:

Recycling:

Recovery of cobalt from spent batteries is becoming more significant.

Methods include mechanical recycling, hydrometallurgical procedures, and pyrometallurgical processes.

Aim to minimize dependency on primary mining and lessen environmental effects.

Reduction in Battery Chemistries:

Research investigates battery chemistries that require less or no cobalt, such as NMC (Nickel

Manganese Cobalt) 811 (high nickel, low cobalt) or LFP (Lithium Iron Phosphate) batteries.

Advances in solid-state batteries and other technologies may further decrease or eliminate the demand for cobalt.

Nickel

Supply Chain:

Major Producers and Regions:

Indonesia, the Philippines, Russia, Canada, and Australia are major nickel producers.

Indonesia and the Philippines are renowned for laterite nickel, whereas Russia and Canada are noted for sulfide nickel deposits.

Demand:

High-Nickel Battery Chemistries:

High nickel concentration in batteries (e.g., NMC 811) boosts energy density and minimizes cobalt reliance.

Demand for high-purity nickel sulfate for battery manufacture is growing.

Challenges:

Environmental and Geopolitical Issues:

Laterite nickel mining is ecologically destructive, including deforestation, soil erosion, and pollution.

Geopolitical conflicts, especially with big suppliers like Russia, may undermine supply stability.

Processing laterite ore is energy-intensive and creates large greenhouse emissions.

Graphite

Natural vs. Synthetic Graphite:

Natural Graphite:

Mined from natural reserves, particularly in China, Brazil, and India.

Requires purification to eliminate contaminants before use in batteries.

Synthetic Graphite:

Produced from petroleum coke by a high-temperature process.

More pricey yet gives more purity and consistency.

Production Processes:

Natural Graphite:

Mining and flotation to separate graphite flakes.

Purification employs chemical or thermal processes to obtain battery-grade purity.

Synthetic Graphite:

High-temperature graphitization method of petroleum coke.

Involves tremendous energy usage yet is produced in very pure and customizable material.

Market Outlook and Future Trends:

Demand Growth:

Driven by EV battery demand, notably for anodes.

Increasing usage in other high-tech applications including fuel cells and nuclear reactors.

Supply Challenges:

Dependence on China for natural graphite creates supply danger.

Environmental laws on mining and processing are tightening, hurting productivity.

Innovation:

Research exploring alternatives like silicon anodes or hybrid composites mixing graphite with other substances to boost performance and lower costs.

Manganese

Uses in EV Batteries:

Role in NMC Cathodes:

Used in conjunction with nickel and cobalt to create NMC cathodes.

Enhances the thermal stability and safety of the battery.

Lower cost compared to cobalt and nickel.

Supply Chain:

Major Producers:

South Africa, Australia, China, and Gabon are key producers.

Mining encompasses open-pit and subterranean processes.

Challenges:

Market Volatility:

Prices are impacted by demand changes in the steel sector, where manganese is also commonly utilized.

Supply chain problems are due to political instability or environmental constraints.

Future Prospects:

Potential for Growth:

Increased usage in EV batteries might raise demand.

Innovations in battery chemistry may boost manganese use and efficiency.

# CHAPTER 3

## Sustainability and Ethical Sourcing

## Environmental Impact of Mining and Refining

The Application of sustainability concepts to mining is inherently hard since mining involves the process of taking and utilizing a restricted resource. However, consideration of sustainability – meeting present needs without compromising the needs of future generations – is increasingly being incorporated into mine development and operation as demand for minerals and products of mining such as metals and fuel and non-fuel minerals and the environmental impacts associated with minerals extraction activities continue to increase. This study presents a survey of current research literature and opinion on sustainability in the mining of non-fuel minerals.

A common sustainable mining framework is focused on mitigating the environmental consequences of mining. Strategies for evaluating the sustainability of mining operations involve measuring, monitoring, and striving to improve several environmental performance criteria, and they are used to evaluate whether a mining operation is sustainable. The major measures for environmental sustainability in mining relate to efficiency in resource use, limiting land disturbance, pollution reduction, as well as closure and reclamation of exhausted mine areas.

Another sustainable mining framework transitions from the emphasis on the environmental footprint of mining operations to responsible management of non-fuel mineral resources throughout their entire life cycle, including the use phase and end of life, with attendant implications for reducing the quantity of mined material and preserving reserves for future generations. This study addresses the shift to a larger framework that incorporates the complete mineral life cycle. We propose that the assessment of sustainability in mining should encompass a systems view of mined materials in society, emphasizing existing environmental sustainability metrics from mining operations as well as including "circularity" or "life cycle" metrics to assess the sustainability of production and extraction for the long term.

The mining and refining procedures for essential elements used in electric vehicle (EV) batteries have severe environmental implications. Understanding these implications is vital for designing sustainable practices and avoiding negative outcomes.

Overview of Key Materials

Lithium

Sources: Primarily collected from brine pools in South America (Chile, Argentina) and from hard rock (spodumene) in Australia.

Uses: Crucial for lithium-ion batteries used in EVs because to its high energy density.

Cobalt

Sources: Mainly mined in the Democratic Republic of Congo (DRC).

Uses: Stabilizes battery cathodes, boosting energy density and lifetime.

Nickel

Sources: Found in nations including Indonesia, the Philippines, and Russia.

Uses: Enhances energy density and storage capacity in battery cathodes.

Graphite

Sources: Predominantly supplied by China.

Uses: Essential for anodes in lithium-ion batteries.

Manganese

Sources: Mined in South Africa, Australia, and Gabon.

Uses: Provides thermal stability and is utilized in several battery chemistries.

## 2. Environmental Impacts of Mining

### Lithium Mining

Water Usage: Extracting lithium from brine pools entails pumping large volumes of water, leading to water depletion in dry locations. This may disrupt local agricultural and water supply.

Chemical Pollution: The extraction process may result in the release of toxic chemicals into the environment, damaging soil and water resources.

### Cobalt Mining

Soil and Water Pollution: Mining operations may lead to heavy metal pollution of soil and water bodies, causing health dangers to local residents and ecosystems.

Deforestation: Cobalt mining frequently requires destroying huge sections of forest, resulting to habitat damage and loss of biodiversity.

Nickel Mining

Greenhouse Gas Emissions: Processing nickel ores, especially laterite ores, is energy-intensive and causes large greenhouse gas emissions.

Acid Mine Drainage: Sulfide ores may cause acid mine drainage, which contaminates water sources with sulfuric acid and dissolved heavy metals.

Graphite Mining

Air and Water Pollution: Graphite mining and processing may create dust and emit chemicals that harm the air and water.

Land Degradation: Mining activities may contribute to soil erosion, land subsidence, and landscape alteration.

Manganese Mining

Water Pollution: Mining and refining manganese may bring toxins into water bodies, impacting aquatic life and water quality.

Habitat Disruption: Mining operations disturb local ecosystems and animal habitats, resulting to biodiversity loss.

3. Environmental Impacts of Refining

Energy Consumption

Substantial Energy Demand: Refining operations for lithium, cobalt, nickel, and other minerals are energy-intensive, leading to substantial carbon emissions, particularly when driven by fossil fuels.

Chemical Waste

Poisonous Byproducts: Refining includes the use of chemicals that create poisonous byproducts. Improper disposal or inadvertent discharge of these compounds may create environmental contamination and health concerns.

Water Usage

Water-Intensive Processes: Many refining procedures demand huge volumes of water for cooling, cleaning, and chemical reactions, worsening water shortage difficulties in certain places.

4. Mitigating Environmental Impacts

Sustainable Mining Practices

Water Management: Implementing effective water use and recycling procedures in mining operations may decrease water consumption and limit effects on local water resources.

Rehabilitation and Reclamation: Restoring mined land via reforestation, soil stabilization, and habitat restoration helps reduce environmental harm and encourages ecosystem recovery.

Monitoring and Regulation: Enforcing strong environmental legislation and monitoring programs assures compliance with sustainable mining operations and lowers pollution.

Green Refining Technologies

Renewable Energy Integration: Utilizing renewable energy sources (solar, wind, hydro) in refining operations minimizes greenhouse gas emissions and lessens dependency on fossil fuels.

Closed-Loop Systems: Implementing closed-loop systems for chemical usage and waste management decreases harmful waste creation and encourages recycling and reuse of resources.

Advanced Filtration and Treatment: Employing advanced filtration and water treatment technologies decreases the environmental effect of water consumption and assures the safe disposal of industrial effluents.

Ethical Sourcing Initiatives

Supply Chain Transparency: Enhancing transparency in the supply chain using traceability technology (blockchain, RFID) helps guarantee that raw materials are supplied lawfully and ethically.

Third-Party Certification: Certifications such as Fairtrade and Responsible Cobalt Initiative (RCI) ensure that resources are obtained in line with ethical and environmental standards.

Stakeholder Collaboration: Collaboration among governments, NGOs, and industry stakeholders supports the creation of best practices, standards, and regulations that promote sustainable and ethical sourcing.

5. Case Studies and Examples

Sustainable Lithium Extraction in Chile

Water-Saving technology: Companies are investigating technology that minimizes water usage in lithium extraction from brine pools, helping to protect local water resources.

Cobalt Mining in the DRC

Ethical Mining Programs: Initiatives like the Fair Cobalt Alliance attempt to improve mining processes, assure fair salaries, and eradicate child labor in cobalt mining areas.

Nickel Refining Innovations

Carbon Reduction Strategies: Some nickel refining businesses are investing in carbon capture and storage (CCS) systems to minimize greenhouse gas emissions from processing activities.

Graphite Mining in China

Environmental restrictions: The Chinese government has established stronger environmental restrictions for graphite mining, resulting to the closure of non-compliant enterprises and development of cleaner technology.

Manganese Mining in South Africa

Community Engagement: Mining firms are cooperating with local communities to resolve environmental issues, give job opportunities, and invest in community development initiatives.

The environmental effect of mining and refining critical elements for EV batteries offers substantial issues that must be addressed to guarantee sustainable and ethical sourcing. By embracing innovative technology, sustainable practices, and ethical activities, the sector may offset these consequences and contribute to a more sustainable future. Through thorough knowledge and continual development, stakeholders may work together to promote responsible resource management and assist the worldwide transition to electric transportation.

# CHAPTER 4

## Supply Chain Dynamics and Challenges

### Overview of the EV Battery Supply Chain

The supply chain for electric vehicle (EV) batteries is complex and global, involving multiple stages from raw material extraction to battery manufacturing, and ultimately, recycling. Understanding this supply chain is crucial for identifying challenges and opportunities in the EV industry.

### 1. Raw Material Extraction

#### Key Raw Materials

Lithium: Essential for lithium-ion batteries, lithium is primarily extracted from brine pools in countries like Chile and Argentina or from hard rock deposits in Australia.

Cobalt: Used to stabilize battery cathodes, cobalt is mainly sourced from the Democratic Republic of Congo (DRC), which accounts for over 60% of global production.

Nickel: Increases energy density and storage capacity of batteries, with major suppliers including Indonesia, the Philippines, and Russia.

Graphite: Used for anodes, with China being the dominant producer and supplier.

Manganese: Provides thermal stability and is primarily mined in South Africa, Australia, and Gabon.

Environmental and Ethical Concerns

Mining Impacts: Extraction processes can lead to significant environmental degradation, including water pollution, habitat destruction, and increased carbon emissions.

Human Rights Issues: Particularly in cobalt mining, there are concerns about child labor, poor working conditions, and exploitation, especially in the DRC.

2. Material Processing and Refining

Conversion and Purification

Lithium Processing: Extracted lithium undergoes several chemical processes to produce lithium carbonate or lithium hydroxide, essential for battery production.

Cobalt Refining: Involves purifying cobalt to remove impurities and produce battery-grade cobalt sulfate.

Nickel Refining: Nickel laterite and sulfide ores are processed to produce nickel sulfate, used in battery manufacturing.

Geopolitical Factors

Concentration of Processing Facilities: A significant portion of material processing and refining is concentrated in China, leading to geopolitical risks and potential supply chain disruptions.

Trade Policies and Tariffs: International trade policies, tariffs, and export restrictions can impact the availability and cost of raw materials.

3. Battery Component Manufacturing

Cathodes and Anodes

Cathode Materials: Commonly used materials include Lithium Nickel Manganese Cobalt Oxide (NMC) and Lithium Iron Phosphate (LFP). The choice of cathode material affects the battery's energy density, cost, and safety.

Materials used for the anode are mostly composed of graphite, however there is current research into silicon-based anodes to achieve better energy density.

In addition to electrolytes, separators

Separators: Thin membranes that avoid short circuits by keeping the cathode and anode apart while allowing ionic passage. Materials like as polyethylene and polypropylene are used in its construction.

Electrolytes: Liquid solutions allowing ion transfer between the cathode and anode. in most cases, are composed of lithium salts that have been dissolved in organic solvents.

Manufacturing Processes

Mixing and Coating: After the active ingredients have been combined with binders and conductive agents, they are coated onto metal foils (aluminum for cathodes and copper for anodes).

In order to construct cells, coated foils are wrapped or layered in order to create cells. These cells are then filled with electrolytes and sealed again.

The formation and aging process involves the charging and discharging cycles that newly produced cells go through in order to stabilize and improve their performance before they are finally combined into battery packs.

4. Battery Pack Assembly

Assembly of the Modules and Packs

Modules are a grouping of several cells that incorporate safety elements such as heat management and electrical connections. Modules are designed to accommodate multiple cells.

In the process of assembling modules into packs, cooling systems, battery management systems (BMS), and structural components are included in the packs.

Quality Control and Testing

Performance Testing: Ensures that cells, modules, and packs fulfill specified performance parameters for capacity, energy density, and cycle life.

Safety Testing: Includes testing for thermal runaway, short-circuit resistance, and impact resilience to assure safety under diverse situations.

5. Distribution and Logistics

Supply Chain Management

Transportation: Involves moving raw materials, components, and completed battery packs around the world, frequently needing specialized handling and packing to minimize damage and assure safety.

Inventory Management: Balancing supply and demand to maintain ideal inventory levels, decrease costs, and prevent production delays.

Challenges and Risks

Logistical Complexities: Coordinating a worldwide supply chain with various stakeholders and significant lead times may be tough.

Disruptions: Natural catastrophes, geopolitical conflicts, and pandemics may interrupt supply chains and lead to material shortages or delays.

6. Recycling and End-of-Life Management

Battery Recycling Processes

Collection and Sorting: Used batteries are gathered and sorted depending on chemistry and condition.

Dismantling: Batteries are disassembled into constituent components for subsequent processing.

Material Recovery: Processes like pyrometallurgy, hydrometallurgy, and direct recycling are used to recover precious materials like lithium, cobalt, and nickel.

Circular Economy

Second-Life Applications: Batteries that no longer match EV performance criteria may be recycled for less demanding applications, such as stationary energy storage.

Sustainability Goals: Recycling and second-life uses contribute to a circular economy, lowering the demand for new raw materials and limiting environmental effects.

The EV battery supply chain is a sophisticated and internationally integrated network, covering raw material extraction, processing, component production, assembly, distribution, and recycling. Each stage brings distinct difficulties and possibilities, impacted by technical improvements, geopolitical dynamics, environmental concerns, and regulatory frameworks. A strong and resilient supply chain is vital for the sustainable expansion of the EV sector, assuring the availability of high-quality batteries while resolving ethical, environmental, and logistical problems

# CHAPTER 5

## Charging Infrastructure and Grid Integration

### Overview of EV Charging Infrastructure

### Types of Charging Stations

Level 1 Charging: Uses a typical 120-volt AC household outlet, giving the slowest charging rate suited for overnight charging at home or at businesses where cars are parked for lengthy durations. Typically adds roughly 2-5 miles of range every hour of charging.

Level 2 Charging: Utilizes a 240-volt AC source and produces greater power levels than Level 1 chargers. Commonly seen in home settings, offices, and public spaces like shopping malls. Level 2 chargers may contribute around 10-30 miles of range per hour of charging, depending on the car and charger specs.

DC Fast Charging (DCFC): Provides quick charging by supplying direct current (DC) to the vehicle's battery, bypassing the onboard charger. DCFC stations are vital for long-distance travel and high-demand locations, capable of charging EVs to 80% capacity in around 20-30 minutes.

Charging Station Networks

Public Charging Networks: Operated by firms like as ChargePoint, EVgo, and Electrify America, these networks provide a range of chargers, including Level 2 and DCFC stations, at public sites such as rest areas, retail malls, and urban centers. They allow easy charging access for EV users who may not have access to private chargers.

Private Charging Infrastructure: Includes residential charging stations built by homeowners, workplace chargers given by employers, and fleet charging solutions for commercial cars. These home installations facilitate regular EV charging routines and help to minimize dependency on public charging infrastructure.

Charging Connector Standards

North America: CCS (Combined Charging System) is widely used for DC fast charging, merging both AC and DC charging capabilities into a single connection. Tesla automobiles employ a proprietary connection but also supply adapters for CCS compatibility.

Europe: CCS Combo 2 is popular, including Type 2 connections for AC charging and CCS for DC fast charging. This standardization facilitates interoperability across multiple charging networks and promotes consumer comfort throughout European nations.

Asia: CHAdeMO remains common in Japan, offering both AC and DC charging. In China, the GB/T connection standard is implemented alongside CCS, reflecting regional preferences and infrastructural improvements.

Smart Charging Features

Load Management: Smart chargers can dynamically alter charging rates depending on grid demand, energy costs, and car owner preferences. This improvement helps to lessen pressure on the system during peak demand times and lowers total power expenditures for EV users.

Remote Monitoring: Charging stations equipped with remote monitoring capabilities give real-time data on charging status, energy usage, and operating efficiency. This data helps network operators to improve charging station performance, fix problems swiftly, and assure dependable service for EV consumers.

2. Impact on Grid Integration

Grid Capacity and Stability

Peak Demand Management: The widespread adoption of EVs and high-power charging stations might lead to localized grid congestion during peak hours. Smart charging infrastructure helps manage and distribute charging demands intelligently, reducing grid overload and limiting the need for expensive grid modifications.

Grid Resilience: Bi-directional charging technologies, such as Vehicle-to-Grid (V2G) systems, allow EVs to not only consume power but also return surplus energy to the grid. This feature promotes system stability, enables renewable energy integration, and offers backup power during grid disturbances or crises.

Infrastructure Planning and Expansion

Urban vs. Rural Deployment: Urban locations need dense networks of fast chargers to support large numbers of EVs and fulfill the charging demands of urban inhabitants who may not have access to private charging facilities. In contrast, rural areas benefit from strategically situated fast chargers along highways and important travel routes to promote long-distance travel and encourage EV adoption in less densely populated locations.

Interoperability: Standardization of charging protocols and connection types across regions and nations is vital for delivering smooth EV charging experiences for drivers traveling abroad or utilizing multiple charging networks. Regulatory measures and industry cooperation seek to improve interoperability and simplify the EV charging infrastructure for customers.

Policy and Regulatory Considerations

Deployment Incentives: Government incentives, including as subsidies, tax credits, and grants, play a crucial role in promoting private sector investment in charging infrastructure deployment and growth. These incentives assist cover installation costs, boost network expansion, and hasten the transition to electric transportation.

Standardization Initiatives: Regulatory frameworks and industry standards support the adoption of common charging protocols and connection standards, promoting market competition, expanding customer choice, and assuring dependability across varied charging networks.

3. Technological Advances and Future Trends

Ultra-Fast Charging

High-Power Chargers: Technological innovations continue to push the boundaries of charging speed and efficiency, with ultra-fast chargers capable of providing up to 350 kW or more. These high-power chargers considerably cut charging periods, making long-distance EV travel more easy and feasible.

Liquid-Cooled Cables: Enhance charging efficiency by dispersing heat created during high-power DC rapid charging sessions. Liquid-cooled cables assist in maintaining ideal charging temperatures, increasing charging dependability, and prolonging the lifetime of charging infrastructure components.

Wireless Charging

Inductive Charging: Wireless charging technology enables EVs to charge without physical touch using charging pads embedded in parking places, highways, or garage floors. Inductive charging systems simplify the charging process, save wear and tear on physical connections, and facilitate autonomous vehicle movement for exact alignment with charging pads.

Grid-Interactive Charging

Vehicle-to-Grid (V2G) Systems: Bi-directional energy flow allows EVs to store excess power from renewable sources or off-peak hours and return it to the grid during peak demand periods or system crises. V2G systems boost grid stability, assist renewable energy integration, and give financial incentives to EV owners via energy arbitrage.

Battery Swapping

Automated Stations: Battery swapping stations give an alternative to conventional charging by enabling EV users to exchange exhausted batteries for fully charged ones within minutes.

Automated battery swapping reduces EV downtime, addresses range anxiety for commercial fleet operators and supports continuous vehicle operation without waiting for battery recharging.

Electric vehicle charging infrastructure plays a critical role in enabling the widespread adoption and integration of electric mobility solutions. Technological advancements, supporting legislative frameworks, and strategic infrastructure investments are vital for extending charging accessibility, strengthening grid stability, and improving the overall user experience for EV drivers. As the electric vehicle industry continues to expand, establishing strong, interoperable, and grid-integrated charging systems will be important in advancing the transition to sustainable transportation and decreasing carbon emissions worldwide.

# CHAPTER 6

## Solid-State Batteries

### Technology Overview

**Solid-State Electrolytes:** Solid-state batteries replace the liquid or gel electrolytes present in standard lithium-ion batteries with solid electrolytes. These may be ceramic, polymer, or glass-based materials. They provide superior energy density and safety advantages compared to liquid electrolyte batteries owing to their inherent stability and less chance of leakage or fire.

**Advantages:** Solid-state batteries offer increased energy density, perhaps quicker charging times, and improved safety profiles, making them suitable for electric vehicles (EVs) seeking for extended ranges and enhanced performance.

Current Development Status

Challenges: Despite encouraging laboratory findings, scaling up production remains a problem owing to challenges like as manufacturing costs, stability for long-term usage, and obtaining consistent performance across diverse operating circumstances (temperature changes, charge-discharge cycles).

Research Focus: Ongoing research focuses on improving solid-state electrolytes for higher conductivity, addressing interface difficulties between solid electrolytes and electrode materials, and establishing scalable production procedures.

Potential Benefits

Performance: Solid-state batteries might dramatically enhance the driving range of EVs per charge, shorten charging times to minutes rather than hours, and improve operating safety.

Uses: Besides EVs, solid-state batteries have potential uses in consumer electronics, aircraft, and stationary energy storage owing to their small size, high energy density, and better safety.

2. Advancements in Lithium-Ion Technology

High-Nickel Cathodes

Chemistry: High-nickel cathodes like NMC 811 (Nickel Manganese Cobalt oxide with an 8:1:1 ratio) provide better energy density compared to typical cathode formulations. They do this by lowering the cobalt concentration, which is pricey and connected with ethical problems relating to mining techniques.

Performance Characteristics: These cathodes boost the energy storage capacity of lithium-ion batteries, but they confront issues such as thermal stability at high temperatures and probable degradation during charge-discharge cycles.

Future Prospects: Continued research intends to increase the stability and cycle life of high-nickel cathodes while lowering their dependence on cobalt, therefore driving down prices and boosting sustainability.

Silicon Anodes

Capacity Improvement: Silicon offers a large theoretical capacity for lithium-ion storage, greatly surpassing that of graphite normally employed in anodes. Silicon anodes offer much greater energy densities, perhaps tripling the energy storage capacity of present batteries.

Challenges: Silicon endures considerable volume changes during charge-discharge cycles, resulting to mechanical stress and electrode deterioration. Researchers are studying nanostructuring methods and silicon-carbon composites to address these problems.

Research Directions: Advances in silicon anodes might transform battery technology by enabling greater driving ranges and decreasing the total weight of EV batteries.

3. Sustainable Battery Materials

Cobalt Reduction

Importance: Cobalt is a vital component in many lithium-ion batteries, but its mining is related to environmental deterioration and human rights concerns, including child labor in certain locations.

alternate Chemistries: Researchers are exploring alternate cathode chemistries (e.g., NMC 622, NMC 532) with lower cobalt concentrations to solve these problems while preserving battery performance and stability.

Recycling Initiatives: The recycling of lithium-ion batteries is vital for lowering dependency on freshly mined cobalt and other rare elements. Advanced recycling methods strive to recover and utilize cobalt, lithium, and other important metals from used batteries effectively.

Environmental Impact: Sustainable battery materials not only decrease environmental impact but also contribute to the circular economy by reducing resource extraction and supporting the reuse of crucial elements.

4. Charging Infrastructure and Battery Management

Fast Charging

Technology: High-power charging stations capable of providing up to 350 kW are being implemented internationally to cut charging times dramatically. These stations employ cooled cables and specialized power circuitry to preserve battery health during fast charging.

Impact on Battery Life: Fast charging may accelerate battery deterioration owing to increased heat production and mechanical stress on electrode materials. Battery management systems (BMS) play a significant role in reducing these consequences by optimizing charging profiles and regulating temperature conditions.

Battery Management Systems (BMS)

Functions: BMS monitors battery state-of-charge, voltage levels, and temperature to guarantee safe operation and optimal performance. Advanced BMS algorithms allow predictive maintenance and real-time modifications to enhance battery longevity and efficiency.

Technological Advances: Integration of artificial intelligence (AI) and machine learning (ML) algorithms in BMS enables for adaptive charging procedures adapted to individual battery characteristics and ambient circumstances.

5. Integration with Renewable Energy

Vehicle-to-Grid (V2G) Systems

Concept: V2G technology enables EVs to store extra power from renewable sources (e.g., solar or wind) and discharge it back to the grid during peak demand times. This bi-directional energy transfer maintains grid stability and increases the value proposition of EVs beyond mobility.

Technical Requirements: Standards and protocols are being created to assure compatibility between V2G-capable cars and grid infrastructure, addressing difficulties such as power quality and interoperability.

Market Adoption: Pilot projects and regulatory incentives are encouraging the adoption of V2G systems, with potential advantages for both EV owners and grid operators in terms of energy cost reductions and resilience.

Energy Storage Solutions

Dual Purpose: EV batteries may function as decentralized energy storage devices, helping to grid stability by smoothing out swings in renewable energy supply and providing backup power during outages.

Technological Advancements: Advances in energy management systems allow efficient charging and discharging schedules depending on power pricing, grid demand, and customer preferences.

# 6. Market Trends and Adoption

## Cost Reduction

**Economies of Scale:** Increased manufacturing volumes and technical improvements are bringing down the cost of EV batteries, making electric cars more accessible for customers.

**Material Innovations:** Research into novel battery chemistries and production techniques attempts to lessen the dependence on pricey materials like cobalt and enhance overall battery efficiency.

**Policy Incentives:** Government subsidies, tax credits, and regulatory requirements encourage manufacturers to invest in electric car manufacturing and battery technologies, speeding cost reductions and market acceptance.

Global Expansion

Regional Dynamics: Variations in EV adoption rates and regulatory environments across regions (e.g., Europe, North America, Asia) impact market growth and investment plans for battery production facilities.

Infrastructure Development: Collaborative efforts between governments, manufacturers, and utilities are developing charging networks and enabling the mainstream adoption of electric cars.

7. Regulatory and Policy Landscape

Emission Standards

Regulatory Frameworks: Stricter emissions laws and objectives (e.g., EU $CO_2$ requirements, US CAFE criteria) are encouraging manufacturers to electrify their vehicle fleets and expand the manufacturing of electric cars.

Impact on Industry: Compliance with emissions requirements impacts product planning, investment choices, and technological development within the automobile industry, with substantial consequences for battery demand and production.

Government Incentives

Financial Support: Incentives such as tax credits, rebates, grants, and subsidies minimize the initial costs of electric cars and encourage people to transition from internal combustion engine vehicles.

Policy Stability: Long-term policy frameworks offer confidence for automakers and investors, promoting continual innovation in battery technology and charging infrastructure.

## 8. Challenges and Future Outlook

Technological Hurdles

Energy Density: Improving the energy density of batteries beyond present limitations is vital for improving the driving range of electric cars and boosting their overall efficiency.

Lifecycle Sustainability: Addressing difficulties relating to battery longevity, degradation processes, and recycling technologies to reduce environmental effect and increase resource efficiency.

Emerging Technologies: Exploration of post-lithium battery technologies (e.g., solid-state, lithium-sulfur) and their potential to overcome present constraints in energy density, cost, and environmental sustainability.

Supply Chain Constraints

Raw Material Security: Ensuring a consistent supply of crucial materials (e.g., lithium, cobalt, nickel) necessary for battery manufacture despite geopolitical conflicts, resource depletion concerns, and shifting market pricing.

Circular Economy: Developing closed-loop recycling methods to recover valuable materials from old batteries and minimize dependency on primary resource exploitation.

Consumer Acceptance

Education and Awareness: Addressing customer issues such as range anxiety, charging infrastructure availability, and the overall cost of ownership to boost confidence in electric cars as viable alternatives to conventional combustion engine vehicles.

User Experience: Enhancing the usability and convenience of electric cars via enhanced charging networks, battery performance, and vehicle design to attract a larger customer base and speed up market adoption.

# CHAPTER 7

## Environmental Impacts and Sustainability of EV Batteries

The environmental impact and sustainability of electric vehicle (EV) batteries are essential factors as civilizations migrate towards greener transportation alternatives. This chapter presents an in-depth investigation of the lifecycle analysis, environmental concerns, sustainability efforts, and regulatory frameworks impacting EV battery manufacture, consumption, and disposal.

## Lifecycle Analysis of EV Batteries

### Resource Extraction and Manufacturing

EV batteries depend on critical elements such as lithium, cobalt, nickel, and graphite, each having its own environmental concerns. The extraction of these basic minerals includes mining activities that may lead to land degradation, water pollution, and disturbance of local ecosystems. For instance, lithium mining sometimes takes enormous quantities of water and might impact delicate habitats like salt flats.

Manufacturing Processes

During battery manufacture, substantial energy is used, mostly from fossil fuel sources, adding to greenhouse gas emissions. Efforts are undertaken to increase the energy efficiency of manufacturing processes and move towards renewable energy sources to minimize carbon footprints. Advanced manufacturing processes are also being developed to eliminate waste and maximize material consumption in battery construction.

Use Phase

EVs provide substantial benefits over internal combustion engine vehicles (ICEVs) in terms of decreasing greenhouse gas emissions and improving air quality, particularly when driven by renewable energy sources. However, the total environmental advantages during the usage period rely on the energy mix utilized for charging. Regions with greater renewable energy penetration gain more in terms of carbon emission reductions.

End-of-Life Considerations

As EV batteries near the end of their operating life, efficient disposal and recycling are vital to reduce environmental effects. Recycling programs recover precious minerals such as lithium, cobalt, and nickel, minimizing the need for new mining and protecting natural resources. Effective recycling also mitigates the danger of environmental contamination from hazardous battery components.

Environmental Challenges and Mitigation Strategies

Carbon Footprint Reduction

Reducing the carbon footprint of EV batteries includes tackling emissions across the supply chain, from raw material extraction to manufacture and disposal. Strategies include:

Supply Chain Optimization: Enhancing efficiency and transparency in material procurement and transportation to minimize emissions.

Renewable Energy Adoption: Transitioning towards renewable energy sources in manufacturing plants and during battery charging to decrease lifetime greenhouse gas emissions.

Carbon Collect and Storage (CCS): Exploring ways to collect and store $CO_2$ emissions produced during battery manufacture and manufacturing operations.

Water and Land Use

The mining and processing of battery materials may have considerable water and land usage consequences, especially in locations with sensitive ecosystems. Sustainable practices include:

Water Recycling and Conservation: Implementing water recycling systems and lowering water use in mining and industrial activities.

Land Rehabilitation: Rehabilitating mining areas post-extraction to restore natural ecosystems and avoid long-term environmental harm.

Sustainability Initiatives and Standards

Certification and Standards

Environmental certifications such as ISO 14001 and EMAS guarantee that EV battery manufacturers comply to high environmental requirements and constantly improve their environmental performance. These certifications address elements ranging from waste management and emissions reduction to sustainable sourcing practices.

Circular Economy

Promoting a circular economy strategy entails designing batteries for a lifetime, simplicity of repair, and recycling. Extended Producer Responsibility (EPR) initiatives push producers to accept responsibility for their goods throughout their lives, supporting the reuse and recycling of materials to decrease waste and environmental effects.

Sustainable Sourcing

Ensuring the ethical procurement of raw materials is vital to minimize environmental degradation and social difficulties linked with mining operations. Initiatives concentrate on:

Transparency and Traceability: Tracing the origin of raw resources to guarantee they are supplied responsibly and ethically.

Conflict-Free Sourcing: Eliminating the usage of conflict minerals (e.g., cobalt supplied from places with human rights violations) via certification programs and ethical supply chain management.

Policy and Regulatory Framework

Government Regulations

Governments worldwide are establishing legislation to encourage sustainable practices in EV battery manufacture and consumption. These restrictions include

Emission Standards: Setting emission limits for battery production operations to decrease air pollutants and alleviate environmental consequences.

Waste Management: Establishing criteria for the safe disposal and recycling of EV batteries to reduce hazardous waste and avoid environmental pollution.

Incentive Programs

Financial incentives like as subsidies, grants, and tax credits support the adoption of EVs and the development of sustainable battery technology. These programs support

Research & Development: Funding programs targeted at enhancing battery efficiency, recyclability, and environmental performance.

Infrastructure Development: Investment in charging infrastructure and battery recycling facilities to boost EV adoption and decrease environmental effects.

Industry Collaboration and Innovation

Research and Development

Ongoing research and development activities concentrate on developing battery technology to boost energy density, durability, and sustainability. Innovations include

Next-Generation Materials: Research on alternative materials and battery chemistries to minimize dependency on precious resources and increase recyclability.

Advanced Recycling Technologies: Development of effective and ecologically friendly recycling technologies to recover valuable materials from wasted batteries.

## Collaborative Initiatives

Industry alliances bring together automakers, battery manufacturers, technology suppliers, and research institutes to expedite innovation and solve sustainability concerns. These projects foster information exchange, best practices, and combined investments in sustainable transportation solutions.

## Consumer Awareness and Education

### Environmental Awareness

Educational initiatives enhance public awareness about the environmental advantages of EVs and sustainable transportation options. Key messages include:

lifetime Environmental Impact: Communicating the lifetime environmental advantages of EVs compared to conventional cars.

Consumer Choice: Empowering customers with knowledge on sustainable purchase choices and the environmental credentials of EVs and their batteries.

Case Studies and Best Practices

Global Examples

Case studies illustrate effective tactics and best practices in increasing environmental sustainability in EV battery manufacture and usage:

Tesla's Battery Recycling Program: Overview of Tesla's closed-loop recycling technique to collect and reuse battery materials.

European Union efforts: Examination of EU policies and efforts supporting sustainable battery manufacturing, recycling infrastructure development, and consumer awareness.

# CHAPTER 8

## Sand Performance of EV Batteries

## Battery Safety Standards and Regulations

## Global Standards and Regulations

Electric vehicle (EV) batteries are subject to severe safety standards and regulations internationally to guarantee their safe functioning and to prevent possible dangers connected with high-voltage systems. These regulations are critical to safeguard customers, automobiles, and infrastructure.

UN ECE rules: The United Nations Economic Commission for Europe (UN ECE) develops rules that are globally recognized and accepted by numerous governments. Key laws include UN ECE Regulation 100 and Regulation 134, which specify safety criteria for electric powertrain systems and high-voltage batteries. These rules address factors such as electrical safety, mechanical integrity, crashworthiness, and environmental performance.

National rules: Countries and regions also have their own particular rules adapted to local situations and market needs. For example:

United States: Federal Motor Vehicle Safety criteria (FMVSS) include FMVSS 305, which specifies safety criteria for electric cars, concentrating on topics like battery integrity during collisions, post-crash electrical safety, and insulation monitoring.

European Union: The EU governs EV batteries under the General Safety Regulation (EU) 2019/2144, which requires safety and performance criteria for all motor vehicles, including electric ones. This rule guarantees that EV batteries fulfill severe safety requirements before they can be marketed in the EU market.

Impact on Battery Design and Engineering

Regulatory criteria greatly impact the design and engineering of EV batteries to guarantee they fulfill safety requirements:

Safety measures: Batteries must have several safety measures to avoid thermal runaway, fires, or explosions. These include;

Thermal Management Systems: Active and passive cooling systems to maintain ideal operating temperatures and avoid overheating.

Battery Management Systems (BMS): Sophisticated BMS that monitor cell voltages, temperatures, and currents in real-time to control charge/discharge cycles and avoid overcharging or over-discharging.

Physical Design: Crash-resistant battery casings and structural components to protect battery modules from damage during crashes.

Durability and Reliability: Standards dictate durability testing to verify batteries can survive mechanical stress, vibration, and climatic conditions during the vehicle's lifespan without affecting safety or performance.

Testing and Certification Processes

Ensuring compliance with safety standards includes rigorous testing and certification processes:

Safety Testing: Comprehensive testing techniques include:

Electrical Safety: Insulation resistance, electrical isolation, and protection against short circuits.

Mechanical Safety: Mechanical integrity tests for structural robustness and impact resistance.

Environmental Testing: Performance in severe circumstances such as high and low temperatures, humidity, and altitude.

collision Safety: Assessment of battery integrity and electrical isolation during collision situations.

Certification Requirements: Manufacturers must submit extensive paperwork and test results to regulatory agencies to establish compliance with safety requirements before commercializing EV batteries. This includes conformity assessment methods and frequent audits to assure continuing compliance.

Emerging Standards and Future Developments

As battery technology improves, new standards and regulatory frameworks are being established to handle growing difficulties and opportunities:

Solid-State Batteries: Emerging technologies like solid-state batteries offer potential improvements in safety and energy density. However, they need additional safety regulations owing to differing material qualities and operating characteristics compared to typical lithium-ion batteries.

Fast-Charging Protocols: Standards are developed to handle the issues of fast charging, including heat production, battery deterioration, and the influence on total battery longevity. Ensuring safety during fast charging is critical to avoid thermal runaway and preserve battery life.

Case Studies and Implementation

Tesla Inc.: Tesla has been at the forefront of implementing strict safety standards throughout its battery production processes and car designs. Their methodology includes;

Advanced Battery Design: Incorporation of numerous levels of safety devices inside battery packs to guarantee strong functioning under varied situations.

Continuous Improvement: Iterative design upgrades and improvements based on real-world data and safety event studies.

Global conformity: Ensuring conformity with international and regional safety standards to promote global market access and customer trust.

sector Collaboration: Collaborative initiatives within the automobile sector seek to standardize worldwide EV battery safety standards. This involves involvement in international standards organizations and forums to unify regulatory requirements and expedite certification procedures.

Future Outlook

Looking forward, the future of EV battery safety and performance will be defined by continued technology improvements and regulatory developments:

Regulatory Trends: Anticipated trends include tougher safety criteria for next-generation battery technology, more attention on environmental sustainability, and laws addressing battery recycling and disposal.

Technological Integration: Advancements in battery management systems, artificial intelligence (AI), and predictive analytics will play a crucial role in increasing safety measures and maximizing battery performance throughout its lifespan.

In conclusion, full adherence to safety regulations and rigorous testing techniques are vital to assuring the safe and dependable functioning of EV batteries. Manufacturers, regulatory agencies, and industry stakeholders must continue to cooperate and innovate to overcome future difficulties and encourage the mainstream adoption of electric cars.

# CHAPTER 9

## Economic and Market Dynamics of EV Batteries

### Global Market Overview

### Market Size and Growth Trends

The electric vehicle (EV) industry has undergone significant expansion driven by environmental legislation, technology improvements, and increasing customer preferences toward sustainable transportation alternatives. Global sales of electric cars have been continuously growing, with a considerable increase predicted in the next years. As of 2022, EV sales topped 6 million units yearly, showing a solid need for EV batteries.

### Regional Market Variations

The adoption of EVs varies among locations and is affected by variables such as government policy, infrastructural development, and consumer awareness. Europe has emerged as a pioneer in EV adoption, backed by rigorous pollution requirements and incentives for EV sales. China is a leading market owing to government backing and aggressive plans for electric car implementation. In North America, measures like zero-emission car requirements and tax incentives have boosted EV adoption rates.

Supply Chain Management and Raw Materials

Critical Components and Raw Materials

The supply chain for EV batteries is complicated and contains multiple important components, including lithium-ion cells, electrodes, electrolytes, and case materials. Key raw materials such as lithium, cobalt, nickel, and graphite are obtained internationally, with supply chain resilience becoming more crucial owing to geopolitical issues and market volatility.

Challenges and Opportunities

Material Sourcing: Securing a steady and ethical supply of raw materials is vital. The industry is researching options to lessen reliance on rare minerals like cobalt and diversify sources of supply.

Environmental Impact: Mining and processing of raw materials cause environmental problems. Efforts are undertaken to increase sustainability via responsible sourcing procedures and recycling programs.

Manufacturing Trends and Innovations

Production Strategies

The introduction of gigafactories—large-scale manufacturing facilities specialized to EV battery production—has been important in fulfilling increased demand and cutting manufacturing costs. Companies like Tesla, CATL, and LG Energy Solution have invested extensively in gigafactory expansions to gain economies of scale and boost production efficiency.

Automation and Industry 4.0 Technologies

Automation technologies, including robots and artificial intelligence (AI), are changing EV battery production. Automated procedures enhance accuracy, minimize labor costs, and maintain uniform quality control throughout production lines. Industry 4.0 concepts allow real-time data monitoring and predictive maintenance, boosting operational efficiency and decreasing downtime.

Cost Drivers and Economics of EV Batteries

Factors Influencing Costs

The cost of EV batteries is a significant driver of car affordability and market competitiveness. Key elements determining prices include:

Material Costs: Fluctuations in raw material prices, notably lithium, cobalt, and nickel, significantly affec battery manufacturing costs.

Economies of Scale: Increased production quantities in gigafactorie contribute to economies of scale, bringing down per-unit costs over time.

Technological Advancements: Innovations in battery chemistry and production techniques lead to cost reductions and performance increases.

Pricing Strategies and Market Competition

Competitive Landscape

The EV battery industry is characterized by fierce rivalry among prominent manufacturers contending for market dominance and technical leadership. Pricing methods are impacted by things such as:

Cost Reduction Goals: Strategies to cut manufacturing costs and transmit savings to customers via competitive pricing.

Differentiation: Companies distinguish their goods via better battery technology, extended warranties, and value-added services like battery recycling programs.

Market Segmentation: Targeting distinct market segments (e.g., passenger cars, commercial fleets) with specialized battery solutions to fulfill varying client demands.

Investment and Funding Landscape

Capital Investments

Investment in EV battery technology encompasses venture capital financing, government grants, and strategic alliances targeted at driving research and increasing manufacturing capacity. Major investments include:

Research and Development: Funding for R&D activities focusing on boosting battery performance, safety, and sustainability.

Infrastructure Development: Investments in charging infrastructure and battery recycling facilities to encourage EV adoption and ecosystem expansion.

Strategic Alliances: Collaborations between automakers, battery producers, and technology businesses to utilize synergies and promote industry breakthroughs.

Consumer Adoption and Market Drivers

Market Influencers

Consumer adoption of EVs is impacted by numerous variables that determine market dynamics:

Regulatory Incentives: Government subsidies, tax incentives, and emission laws that encourage EV adoption by decreasing upfront costs and operational expenditures.

Technological Advancements: Improvements in battery range, charging infrastructure, and vehicle performance boost customer trust in EVs.

Environmental knowledge: Growing knowledge of environmental advantages, including decreased greenhouse gas emissions and lower air pollution levels, increases customer preference towards electric cars.

Industry Outlook and Emerging Trends

Future Projections

The future of EV batteries is set for considerable development and innovation:

Technological Advancements: Anticipated advances in battery chemistry, solid-state batteries, and fast-charging technologies to boost performance and dependability.

Market Expansion: Expansion into new markets such as electric vehicles, buses, and energy storage systems to diversify income streams and seize growing possibilities.

Policy and Regulatory Landscape: Continued development of regulatory frameworks to address sustainability, recycling, and circular economy concepts in battery manufacture and end-of-life management.

Case Studies and Strategic Insights

Real-World Applications

Case studies illustrate successful tactics and advances in EV battery deployment:

Tesla Inc.: Pioneering improvements in battery technology and vertical integration tactics to achieve cost leadership and product differentiation.

CATL and LG Energy Solution: Global growth and strategic alliances to boost market position and fulfill various client requests.

Government activities: Examples of government activities encouraging EV adoption via infrastructure expenditures, research funding, and policy incentives.

www.ingramcontent.com/pod-product-compliance
Lightning Source LLC
Chambersburg PA
CBHW071931210526
45479CB00002B/630